Finding Gold in Washington State: 2005-6 Edition

by

Sean T. Taeschner, M.Ed.

authorHOUSE™

1663 LIBERTY DRIVE, SUITE 200
BLOOMINGTON, INDIANA 47403
(800) 839-8640
WWW.AUTHORHOUSE.COM

© 2005 Sean T. Taeschner, M.Ed. All Rights Reserved.

No part of this book may be reproduced, stored in a retrieval system, or transmitted by any means without the written permission of the author.

First published by AuthorHouse 06/13/05

ISBN: 1-4208-5568-9 (sc)

*Printed in the United States of America
Bloomington, Indiana*

This book is printed on acid-free paper.

Table of Contents

Preface vii

Introduction ix

Section 1 Gold Separation 1

Section 2 Black Sand Separation.. 5

Section 3 Gold and Silver Recycling 9

Section 4 Melting Down Gold... 11

Section 5 *Aqua regia* ("The Royal Water") 13

Section 6 Of Life, Liberty, and the Protection of Fish Eggs . 15

Section 7 Customer Comments (Book Reviews) 21

Section 8 Clubs One Can Join... 27

Section 9 Equipment 35

Section 10 State of Washington Contact Information 45

Section 11 Where To Find Gold In Washington State
    (Author's Favorite Holes) 47
Section 12 Unit & Lesson Plans for Teachers 57
Section 13 Disclaimer 89
Section 14 Questions? 91
Section 15 Acknowledgements 93

Preface

This is a viewable and printable e-book functioning as an informational guide for hobbyists and serious recreational gold prospectors in Washington State.

The guide answers the questions of WHO/WHAT/WHERE/WHEN/WHY & HOW to find, recover, refine and profit from gold while recreational prospecting.

Also included are gold prospecting suppliers and clubs one can join as well as state rules and contact addresses and phone numbers for state resource guides and rules.

The author has also included the hottest areas he has successfully prospected.

His goal is to enrich every person with the ability to be self-supporting through recreational gold prospecting.

Gone Prospecting © 2005 by Sean T. Taeschner, M.Ed.

Introduction

More gold was located and recovered in the United States of America between 1930 and 1940 than at any other time in American history, including the gold rushes of 1849 at Sutter's Mill in California and the Klondike Gold Rush of the 1890s in the Alaskan Yukon Territories.

What caused this massive rush for gold prior to World War Two? The cause was the Great Depression.

With the hurtful recession of the new 21st. Century there is no doubt that this record will be broken once more as Americans lose their corporate jobs and seek to self-empower themselves through recreational gold prospecting.

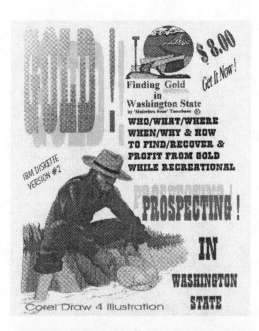

Monica Finds Some Golden Granola at Green River in 2001

Section 1
Gold Separation

Gold is usually found in streams and rivers and deposits itself in cracks and fissures in bedrock, false bedrocks (clays and shales), whirlpool edges, on the downstream-side of rocks, and in moss and grasses on the water's edge.

There is one absolute given in hunting for gold dust or nuggets: where there is black sand you will find gold, platinum, silver, garnet, industrial-grade diamond, bullets, coins and scrap iron. If any of these items show up in your gold pan then you have hit a richly concentrated area. There is no other area except against a slab's base of bedrock sticking right up out of the water where one can find such rich concentrates in black sand!

It is highly recommended that you save all of the black sand that you find. It can be saved up over the year in 5 gallon buckets and sold to steel mills for approximately $1500.00/ton. Black sand is made up of hematite as well as other minerals of gold and silver and platinum too small to see with the naked eye.

The goal here is to teach you how to separate the minute minerals from the black sand so that they, too, aren't overlooked. With gold and platinum at nearly $400.00/troy ounce, it pays to reap all you can from black sand, too.

After taking out the larger nuggets of gold from your black sand by using tweezers, you will see small pinhead-sized flakes or grains of gold. The best way to separate it from the sand is to add dishwashing soap to the water in your pan. This allows the fish oils and tars holding the fine grains of gold the ability to become grease-free so they won't float on top of the water and also stick to the black sand. Using warm or hot water works best.

IT IS SUGGESTED THAT YOU CHANGE THE WATER TWO TO THREE TIMES FOR BEST RESULTS.

Now take a magnet and put it inside a plastic-style Ziploc sandwich bag and dip it into the gold pan. The iron filings will adhere to the bag and then you can put them into a bowl of water nearby and remove them by pulling the magnet out of the plastic bag.

The next step is to pan off the lighter-weight hematite that is left behind. Push the gold flakes into a small pile on the pan's bottom after you have poured all water out. Using your thumb, press on the gold and then touch it to water in a small water-filled vial.

Finding Gold in Washington State: 2005-6 Edition

    If you know an insulin diabetic ask them to save the little glass bottles for you. Such bottles work best for saving and viewing your concentrates.

    The gold sinks out rapidly since it is much heavier than lead. Later, you can separate the fine black sand left behind with your gold by drying it out. In a small blue, red or green cereal bowl blow gently so that the fine dust is blown away and you are left with only 24-carat fine grains of gold. These can then be put into your sample vial for storage until sold or melted down by a jeweler or the U.S. Mint. The U.S. Mint will not take any gold that is less than 1 Troy ounce in weight. Do not bother sending it if you do not have that much to ship. Make sure to insure your shipment with the U.S. Postal Service as well.

Sean T. Taeschner, M.Ed.

John Davis at Green River, 2004. How Do I Get the Gold Out Now?

Section 2
Black Sand Separation

Magnetite saved in a glass jar or 5 gallon buckets by old timers was often crushed with a mortar and pestle into a fine powder and placed into an old iron frying pan (skillet). Adding table salt to the sand (about a half cupful) along with water and a good stir helped separate the gold from the black sand. Only one thing was missing... heat!

This is where modern day prospectors have been known to resurrect methods used by the old time prospectors using backyard barbecues, which have come in handy as a poor-man's gold refining oven.

Building a fire in such barbecues with coals that were red to white hot and then placing the skillet on top had been known to be the old timers' method that worked best for gold separation.

Waiting until the water had boiled out and the sand and salt were red-hot and producing smoke allowed the old timers a safe method to avoid the fumes coming out of the concentrates.

THIS METHOD WAS VERY DANGEROUS DUE TO THE CYANIDE AND/OR MERCURY FUMES GIVEN OFF DURING THE SEPARATION PROCESS; WHICH, IF BREATHED, COULD KILL A PERSON INSTANTLY! THEREFORE, THIS PART OF THE SEPARATION WAS DONE OUTDOORS IN A WELL-VENTILATED AREA!

Once the pan of concentrates had turned cherry-red, the old timers used a long 1 inch steel pipe about four feet long as an extension handle (so they wouldn't burn themselves) in order to place the pan in a tub of cool water, which made the pan cool rapidly.

The gold would then separate out from the sand and could be panned to remove the fine gold or platinum. Modern day prospectors and miners are still saving the sand so they can sell it at the local steel mill for $1400.00/ton.

Remember...a ton is about 2,000 pounds...and it would be equivalent to about 20 five-gallon bucket loads of iron concentrates if sold today!

Finding Gold in Washington State: 2005-6 Edition

John Davis Pans Out His Muddy Concentrates, 2004

Section 3
Gold and Silver Recycling

Old gold jewelry was usually recyclable, because it was made up of gold mixed with silver for extra strength. The higher the carat number of gold the softer it would be and the less silver it would contain! Nowadays, most gold jewelry is a mixture of copper and gold alloy.

Modern day prospectors have been known to use a small hammer and steel block to flatten their gold into a ribbon or large flake to be deposited in a glass jar filled with nitric acid.

Leaving the gold in the acid for a couple of days allowed the acid to react with the silver in the gold. This method left pure gold precipitating out and the fluid left behind in the jar was silver nitrate (AGNo3).

Allowing the gold to be washed in warm water and then letting it stand to dry for several days in the jar produced a

Sean T. Taeschner, M.Ed.

fine gold powder that could be sold for $400.00/ounce to jewelers.

Today's large nugget gold sells for $1500.00/troy ounce since it is often used in pendant and watchband jewelry making.

The silver nitrate could be recycled as well by dropping several copper pennies into the solution...letting it sit for a couple of days. This method allowed modern prospectors a concentration of copper nitrate (Cu No3) and pure silver. These prospectors would pour off the copper nitrate and allow the powdered silver to dry for later usage or to place it in storage for sale at $5.00/troy ounce to chemical supply labs.

Section 4
Melting Down Gold

The Great Depression of the 1930s in the United States of America saw another great gold rush since so many people were out of work and desperate to feed their families. More gold was recovered from 1930 to 1940 than during all previous gold rushes in U.S. history! All of our money used to be backed by gold and silver before the U.S. government printed more than we had in storage in national vaults. Today our money is only backed by the faith people here and abroad have in it.

If it were not for the weekend prospector, our nation's money supply would've shut down long ago. By prospecting for gold, you are adding to this nation's wealth and showing the world what has made this country so great and allowed it to industrialize as quickly as it did.

Sean T. Taeschner, M.Ed.

Every year streams and rivers are replenished with gold that has been brought to the surface of the earth by wind, water, fire and ice and can be harvested by many. Floods often replenish most of this 'resalting' of the gold on our earth as well as volcanic eruptions.

These days gold is melted down in small firebrick furnaces containing a graphite crucible. Pouring the gold out of the crucible with long steel tongs into a cone-shaped mould called a 'steel button' allows the gold to cool and create a 'button of gold'. This makes it easier for the refiner to keep from losing it in fine granular amounts.

Nuggets are rarely melted down, though they are considered small works of art and have their own unique signatures. No two nuggets are ever alike and an assayer can tell what river they came from by their unique shapes, usually.

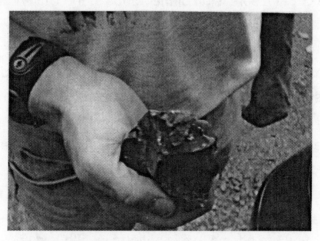

John Davis Holds a White River Mine Sample
Loaded with Fools Gold, 2004

Section 5
Aqua regia ("The Royal Water")

Aqua regia or "Royal Water" received its name centuries ago due to being able to dissolve the "metal of kings," gold, silver and platinum into solution.

Chemists discovered that mixing nitric acid with hydrochloric acid produced an acid so strong it could burn through thick steel plates and keep on going until it was diluted by water. Up until the current day no other acid could successfully melt precious metals from rocks or from other metals. Therefore, the use of *Aqua regia* was both successful for mining and for recycling.

WARNING: The use of acids is VERY DANGEROUS and should only be used in laboratories in safe conditions along with the supervision of licensed, expert chemists.

Sean T. Taeschner, M.Ed.

This author and this book DO NOT CONDONE the use of such methods for leaching precious metals for mining or recycling purposes. This information has been offered for INFORMATIONAL PURPOSES ONLY. Do not try this at home or at work or under any unsafe condition without expert help and guidance.

Websites devoted to this subject appear as public information being shared on the Internet:

http://home.att.net/~numericana/answer/chemistry.htm

http://www.bartleby.com/65/aq/aquaregi.html

Section 6
Of Life, Liberty, and the Protection of Fish Eggs

The debate between environmentalists and miners heated up in the early 1970s in the United States and continues to the present day. Those in the "Green Movement" began labeling those outside of the movement as "not-caring" for the environment. Reflecting on the miners' use of water monitors in the late 1800s (from both photos and movies), which washed down whole mountains and trees, the "Greenies" had valid ammunition to stake their claims. "They" claimed that "miners" were destroying the landscape and its animals and fish by using equipment and methods that were outdated and harmful to the environment.

"They" pointed out, validly, that the use of poisonous mercury by early miners for amalgamating gold concentrates for later smelting caused severe environmental damage. Much of that mercury had washed out of gold pans and had sunken into the sands and gravels of major streams and rivers. (Much of this evidence remained undiscovered until today and is also being cleaned up by dredgers. At $70.00 per one-pound flask it pays to find it and clean it up!) Mercury is poisonous to humans when it touches skin or its fumes are inhaled.

Because of this common knowledge the "Greenies" transferred this poisoning capability onto fish. They were correct. It is illegal to use mercury in mining near streams and rivers. This continues to this day.

The protection of salmon as an endangered species has become another nail in the coffin as miners struggle with the effects of previous mining methodologies. The main argument that has been successfully used by environmentalists and government officials has been that salmon eggs "must be protected at all costs from land developers, miners, farmers and loggers." Years of study by the government and private foundations and universities resulted in a "fix" for the salmon problem.

Beginning in the early 1980s the fix was put into full force. Efforts were made to close logging roads ("due to the silt washing downhill from deforestation into streams and rivers, resulting in the covering of salmon eggs."), keep ATV and motorcycle riders out (for the same reason just noted), miners, and keep farmers from "using poisonous fertilizers that might wash downhill into water used by salmon."

It had been difficult for modern day miners to deny the results of the sins of their predecessors. Therefore, a cooperative effort had to take place between the needs of the "Greenies" and those of miners.

Of Dredging, Cleaning Gravels, and Rotting Fish Eggs

For modern day miners overcoming accusations of ruining spawning habitat would take years. And, sounding the battle cry to make a "difference" against the enemies of water, "Greenies" successfully used the argument that siltation of salmon eggs was a by-product of miners in pursuit of wealth.

The "Greenies" claimed that miners stirred up mud and silt in rivers. The muds and silt covered salmon eggs as they molted. Thus, salmon eggs would die from lack of oxygen and not hatch.

Interestingly enough the "Greenies" failed to mention the yearly devastation caused by flooding from heavy rainfall, as well as the positive aspect dredging caused as gravels were cleaned of oxygen-starving mud.

Normally, fish lay their eggs in gravels in order to allow oxygenated water to circulate around eggs as they mature. However, many eggs actually rot due to gravels becoming plugged with mud, silt and other materials. This causes the loss of millions of salmon eggs each year.

Along came the saviors of the eggs, the dredgers. Each season when dredging is allowed on rivers in our state the spawning gravels are sucked clean of oxygen choking muds and poisonous heavy metals such as lead and mercury. This is a needed benefit to streams and rivers. Additionally, when dredgers create holes near the downriver side of boulders a natural bed is created in which salmon can rest until regaining their strength before heading up river.

Those claiming to be "Environmentalists" have ignored miners in their environmentalism. Miners have become more aware of their own value as environmentalists who remained quiet in the political shadows. Unfortunately, miners have had to come out of their slumber in the past few years. Moreover, they needed to remind the general public that they were becoming as endangered a species as the fish the "Greenies" have vowed to protect. Miners have been robbed of life, liberty, and the pursuit of happiness in exchange for the protection of fish eggs!

Congress affirmed miners' rights to prospect and mine on public lands 133 years ago when they passed the Mining Act of 1872. In that law, mining was seen as a right, not a privilege. Using words as weapons the "Greenies" have attacked the word "recreational" when used in conjunction with prospecting. They hoped to make the serious pursuit of gold mining take on the undervalued / non-serious meaning of a "hobby" activity, instead of the serious personal, scientific, and commercial pursuit of "mining" or "exploring" for gold. Therefore, miners, decided to drop the word "recreational" from mining. Miners believe that mining is not only a privilege and peaceful act of discovery, but also a Congressionally given right under deliberate attack. Words carry power and meaning and can literally change the landscape of America, as we once knew it. It is time to take the country back in both geo and political terms.

Of God, Gold, and Guns and Why They Are Trying to Take Them Away

This author believes that higher powers beyond his simple control have worked behind the scenes for years to slowly discourage and then bankrupt the average citizen from participating in his / her God-given right to worship God, prospect for and profit from gold, and to own and carry firearms for personal use. Never before have the American people seen such an attack on individual and collective rights to use public lands. Never before have we as a people seen the outright attack on the three things that helped build this country in the first place: The right to worship God, search for and mine and own gold, and own guns!

Miners have to turn to the power of Common Sense while mining.

Common sense dictates that miners should perform the following in order to maintain and preserve the rights they currently have:
- Bring your concentrates home to go through in the privacy of your own back yard
- Pan out your concentrates in a large washtub when near a river
- Fill in holes, pack out personal garbage and that of others
- Do not destroy vegetation

- Let the forest ranger know what you are planning to do
- Follow the rules in the Washington State Gold and Fish pamphlet
- Invite Washington State legislators out to pan for gold!

Common sense also dictates that miners consider the following if questioned by authorities while mining:
- Remain calm and don't argue
- Ask if you are being arrested
- If not, walk away peacefully the questioning ends
- If your equipment is confiscated…it is cheap and replaceable, versus a long, expensive court battle

Section 7
Customer Comments
(Book Reviews)

NEWS ARTICLES: June 28, 2002

He Is Still Hoping For Another Gold Rush by Mary Swift of the South County Journal-Kent, WA (Local Section) at the following Web address:

http://www.kingcountyjournal.com/sited/story/html/97115

An Online Book Review by Rodger Petrik of Everett, WA.

Mon. Sept. 01, 2003..."Great information guide on how to find gold in Washington State. Worked the Snohomish river system and found gold. Have now really got the gold bug. I am waiting for the next book on disk to come out for more information and places to go and explore.

Sean T. Taeschner, M.Ed.

Hope he comes out with a book to go rock hounding in Washington State. If you're new or old-hat at gold prospecting, this is the book for you."

Return-Path: <bobnterivert@hotmail.com>
Disposition-Notification-To: "Bob & Teri Vertefeuille" <bobnterivert@hotmail.com>
X-Sender: bobnterivert@hotmail.com
X-Originating-IP: [24.22.248.119]
X-OriginalArrivalTime: 18 Mar 2005 03:45:25.0953 (UTC)
FILETIME=[EBA02310:01C52B6C]
X-UNTD-UBE: -1
X-MimeOLE: Produced By Microsoft MimeOLE V6.00.2800.1441

Sean,

I would strongly recommend your DVD to anyone who wants to learn the basics in prospecting. It was very interesting, and for the beginning prospector, a real easy way to learn. It is more than just grabbing a pan and hitting the river, but much less complicated and involved than one would guess.

I believe that anyone watching your demonstration would immediately "catch the fever". I know when I was finished watching the DVD, I wanted to hit the river. Unfortunately, I have to wait.

Perhaps you could answer a question. Does the permit for prospecting cost anything?

The "spots" I talked to you about in our earlier correspondence are by no means secret. I would love to take anyone who so desired to those very locations and let them see for themselves how rich the gold deposits are.

I am surprised that nobody else had ever worked the area (except the early miners and later the Chinese). As previously mentioned, I have pulled up as many as twenty flakes in one pan. Even my children, who were overly anxious to get to the bottom of their pans found a few flakes. This would be the Buffalo Eddy area on the Snake River south of Asotin on the Washington side of the river.

The other places where I found gold were mostly in Idaho. Although, I did find color on Peshastin Creek and Swauk Creek up on Blewett's Pass near Ellensburg.

I would really be interested in meeting up with you when the season rolls around. I would love to try my luck on this side of the mountains and not have to worry about tangling with poison oak and rattlesnakes.

Hope to hear from you soon. By the way, feel free to quote me any time.

I'll send you the permission to publish form ASAP.

"Stiff Neck" Bob Vertefeuille

Sean T. Taeschner, M.Ed.

John Davis Finds the Motherlode of Iron Pyrite
(Fools Gold) at White River Mine, 2004

Sean- Hello, my name is Charles and we spoke at the Gold Show in Monroe on the 26th. I am the one who bought your DVD at Baker Street Books. I explained to you that I was having trouble with the replacement copy. When I returned home yesterday I cleaned the DVD to make sure the playing surface was clear, and after that it played just fine. I enjoyed watching it very much. It was very instructive and I liked the enthusiasm you put into your demonstrations. Thank you for making this DVD and speaking with me at the show. I found both very educational. I hope to meet you on the Green River some day soon.

Charles zebo@eskimo.com

Sean Showing Off A Lucky Pan of Golden Glitter, 1994 at Horseshoe Bend

Tue, 22 Mar 2005 16:18:05 PST

Hi there, Sean. I hope your week is going well, I just got the second of three shots in my knee. Lots of fun. Here is my narrative.

"I often go out into the wilderness to look for gold in the rivers and streams, or just to go for a hike. Whenever I do I try to be wary of the wildlife that stalks the woods, such as bears or mountain lions. Mountain lions have been known to hide behind things such as trees.

However when I'm looking for gold, I am always more concerned about the two legged animals that like to hide behind a badge.

Sean T. Taeschner, M.Ed.

I'd much rather run into a mountain lion or even a bear!"

How's that? Let me know what you think.

Talk to you later, John.

jdavis98022@yahoo.com

Sluicebox Sean Dreaming of Gold in 2005

Section 8
Clubs One Can Join

Old Diversion Tunnel at Horseshoe Bend on the Sultan River in 1994

Sean T. Taeschner, M.Ed.

<u>National Clubs One Can Join:</u>

GOLD PROSPECTORS ASSOCIATION OF AMERICA
P.O. BOX 891509; TEMECULA, CA 92589- Join for $67.50 on the "Buzzard Special". Tom and Perry Massie, Jake, and Woody own and run the GPAA. These folks have an Alaska trip each year up in Nome, Alaska on the Cripple River. There are also gold mining claims members can work when becoming members of the Lost Dutchman's Mining Association. Lifetime memberships are about $3,500.00. Their phone number is 1-800-233-2207. They also have the Gold Prospector's Show on The Outdoor Channel on cable television.

<u>Washington State Clubs:</u>

BEDROCK PROSPECTORS CLUB
Dwayne Keough...(360) 862-1222...Puyallup, WA... Meetings are held on the third Monday of each month at 7:30 PM at the edge of town at the Washington State Agricultural Station on South Pioneer Street.

BOEING EMPLOYEES PROSPECTORS SOCIETY
Gerald Pollard...(206) 486-0691 / 544-8563...Closed membership to Boeing employees and their families only.

CASCADE TREASURE CLUB
Meets second Sunday each month at the Highland Park Improvement Club...1116 SW Holden Street; Seattle, WA. Meetings begin at 5 PM and visitors are welcome.

Finding Gold in Washington State: 2005-6 Edition

P.O. Box 5208
Wenatchee, WA 98807-5208

North Central Washington Prospectors originally began on December 3, 1977, as was then known as North Central Cascade Miners Association. The dues at that time were $5.00 per year. Records show that the club started with (5) members. Presently, **North Central Washington Prospectors** boasts over 200 members.

The **North Central Washington Prospectors (NCWP)** was formed to assist members in the methods, procedures, and organization of recreational prospecting.

Our intent is to bring together those people with a common interest in all aspects of recreational prospecting and small scale mining, including manufacturers, vendors, dealers of recreational mining equipment, and members of federal, state, and local governments in an effort to promote our common interests.

North Central Washington Prospectors currently meet the third Monday of each month at the Alcoa Aluminum Workers Union Hall located at 180 Rock Island Road in East Wenatchee, WA. The meeting begins at 7:00 p.m.

Membership is open to all persons who wish to participate in and gain knowledge of the recreational prospecting hobby.

As a member of **North Central Washington Prospectors**, you will receive a monthly newsletter, and other information of concern to the recreational prospector.

Members are welcome at all club outings, which could be a one (1) day local metal detecting trip, a weekend gold prospecting outing, or a combination of both.

North Central Washington Prospectors (NCWP) currently have five (5) placer claims on Blewett Pass, Highway 97, in Washington state. Camp sites are available near all the placer claims. These claims are available for use by all members and their immediate family. All the gold that you, the member, finds, is yours to keep!

North Central Washington Prospectors extends an invitation to you to join the club. Current membership dues are as follows: $25.00 a year for singles or a family and $250.00 for a lifetime membership.

President	Vice President	Secretary	Treasurer
Carl Pederson	H. D. Harris	Pat Edwards	Jerry Dillon
509-884-6940	509-662-8427	509-787-1995	509-662-7858
repete@nwinternet.com		patedwards1@juno.com	Ptl2163@aol.com

Sean T. Taeschner, M.Ed.

NORTHWEST PROSPECTORS
Chris Blana...(206) 725-6114...OR (360) 290-6961...Charmaine Jeney, Treasurer/Newsletter editor...(206) 284-1021 in Seattle.

NORTHWEST UNDERGROUND EXPLORATIONS
"Discovering Washington's Historic Mines" Volumes I & II by Co-Author, Vic Pisoni

Contact Information:
Vic Pisoni, Moderator of Northwest Underground Explorations
(206) 722-4238
4215 50th Ave. S.
Seattle, WA 98118-1425
Tunnelhound@Yahoo.com

February 28, 2005

Hi Sean,

Here is our website: NWUNDERGROUND@yahoo.com.

We have 53 members as of this date. Many of these folks are good references in the fields of interest you and I are seeking. We give reports on our most recent mine related hike and research outings. Dates, time, and location for a pre-trip meeting are posted for some of these treks. Some of the explorations are to surface sites and others are of the underground/tunnel workings kind.

Because of the varied personalities and fields of experience involved, NWUNDERGROUND makes for a consolidated core of information to draw on. I know it has saved me much time and effort on some of my data gathering. Chris has included links to other mine information, and one or two very good mining and hiking photo web pages. From time to time we come up with some really fine old historic photos that are open to our member's use.

See you next time,

Victor "Tunnelhound"

http://finance.groups.yahoo.com/group/ NWUNDERGROUND
"This is a loose-knit group of people dedicated to preserving Washington State's mining history through research and exploration and publication. We have been exploring old mines for many years and the towns that once stood nearby. We explore year-round and love what we do. This site was set up to share with people our adventures of exploring old mines and ghost towns all over the State of Washington. So, we hope you enjoy. So, stop in and take a look. (DISCLAIMER) WARNING: Entering abandoned mines is dangerous and could cause harm or death, and should not be attempted by inexperienced people. We do not encourage entering any mine for any reason whatsoever. By joining this site you accept any responsibility for any liability incurred by you or your friends or family or guests from the use of any

Sean T. Taeschner, M.Ed.

information contained herein. Also, we do not encourage or condone trespass on private or otherwise restricted properties without express prior and proper permission from the rightful owners of said property. Notice this website is intended for entertainment purposes only."

*REPRINTED by permission of Vic Pisoni for "Finding Gold in Washington State, 2005-6 Edition by "Sluicebox Sean" T. Taeschner, M.Ed.

OREGON GOLD TRIPS
PRESENTS...

GUIDED/OUTFITTED GOLD MINING TRIPS!

Aerial view above

Email: golddust@oregongoldtrip.com
Website: www.oregongoldtrip.com
TOLL FREE (877) 672-8877

2005 TRIP SCHEDULE!

May	14,15,16
May (Memorial)	27,28,29,30
June	10,11,12
June	24,25,26
Sept. (Labor Day)	2,3,4,5
Sept.	23,24,25
Oct.	7,8,9
Oct.	21,22,23

All our trips are limited to 12 guests.

Transportation is available from Portland, Oregon.

$450/person 3 days
$550/person 4 days

OREGON GOLD TRIPS
P.O. Box 285
Grants Pass, OR. 97528
Toll Free (877) 672-8877
website: www.oregongoldtrip.com
email: golddust@oregongoldtrip.com

RESOURCES COALITION

"Hi Sean.

Thanks for your video. I will look at it this weekend and here is the one I put together for our rally in Oroville, WA. I will e-mail you and sure would like information on printing and duplicating services you are using for your DVD.

Also included are membership forms for Resources Coalition, which you can send back to my return P.O. Box address.

My home phone is 253-863-1709 if you would like calling most evenings. Also, during the day my cell phone number is 425-343-5373. We are reasonably close. I live over in Bonney Lake.

Thanks.

Mark Erickson"

P.S. My e-mail is Inlink@Hotmail.com

*NOTE: This information was reprinted by permission

The 57 minute DVD they put out was excellent. I would recommend it to anyone interested in the yearly rallies held by this club of dredgers and small-scale miners. They are actively working to get the lawmakers in Olympia interested

Sean T. Taeschner, M.Ed.

in attending their rallies in order to see that the miners are also environmentalists. They are to be commended on their efforts.

WASHINGTON PROSPECTORS MINING ASSOCIATION
PMB 1193
10002 Aurora Avenue North #36
Seattle, WA 98133
(206) 784-6039
http://www.washingtonprospectors.org

Section 9
Equipment

Finding Gold with Sluicebox Sean on DVD

Sean T. Taeschner, M.Ed.

*All prices include retail price of $19.95 per copy, 8.9% sales tax on the subtotal of videos ordered, and $2.00 shipping per video.

1 copy= $23.72
2 copies= $47.45
3 copies= $71.18
4 copies= $90.90

Mail purchase order with payment to:
Sean T. Taeschner, M. Ed.
30846 229 PL SE
Black Diamond, WA 98010 USA

Questions: STeshner@Juno.com or (360) 886-1262

Finding Gold in Washington State: 2005-6 Edition

Sean T. Taeschner, M.Ed.

THE HONCOOP HIGHBANKER COMPANY
2004 PRODUCT CATALOG

Products for the gold mining enthusiast.

Featuring
- The Honcoop Highbankers
- Stream Sluice
- Dredge Combo's
- Pumps
- Additional Accessories
- Build-it Yourself Plans

852 NW 1st Ave. #2, Boca Raton FL 33432
Toll-Free 1(888) 988-2122
www.honcoophighbanker.com

 **Jade Drive Rock Shop**

120 E JADE DRIVE
SHELTON, WA. 98584

Phone: 360 426-2327
800 820-3612
Email: jrs@jaderockshop.com
www.jaderockshop.com

Betty & Russ Nation
We carry a complete line of lapidary equipment & supplies.

STORE HOURS:
Mon. thru Sat.
10:00 A.M.- 6.00 P.M.

OKANOGAN PROSPECTING AND MINING SUPPLY
P.O. BOX 47; Riverside, WA 98849 (360) 826-4653. Send for a free brochure.

NORTHWEST TREASURE SUPPLY
Lee Ayerf...P.O. BOX 52802;Bellevue, WA 98015-2802...(425) 881-7340 / Ordering # 1-800-845-5258...Open 8 AM-5 PM M-F and SAT 9 AM-3 PM....Selling Keene dredges and supplies; Fisher pipe locators; Garrett Metal Detectors... For hobbies and security.

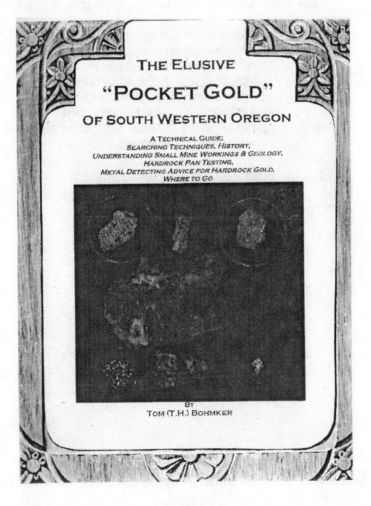

Sean T. Taeschner, M.Ed.

THE ELUSIVE "POCKET GOLD" OF SOUTH WESTERN OREGON

AN ADVANCED TECHNICAL GUIDE BOOK:

- SEARCHING TECHNIQUES
- UNDERSTANDING SMALL MINES WORKINGS & GEOLOGY
- HARD ROCK PAN TESTING
- METAL DETECTING FOR POCKET GOLD
- HISTORY OF POCKET MINING
- HOW TO INTERPRET RESEARCH MATERIALS
- WHERE TO GO

BY: TOM (T. H.) BOHMKER

PUBLISHED BY:

CASCADE MOUNTAINS GOLD
P.O. BOX 33
INDEPENDENCE, OR 97351

PHONE (503) 606-9895

θ Copy right: March 2002
θ Copyright June 2004
Revised Third Edition: θ Copyright Feb 2005

Finding Gold in Washington State: 2005-6 Edition

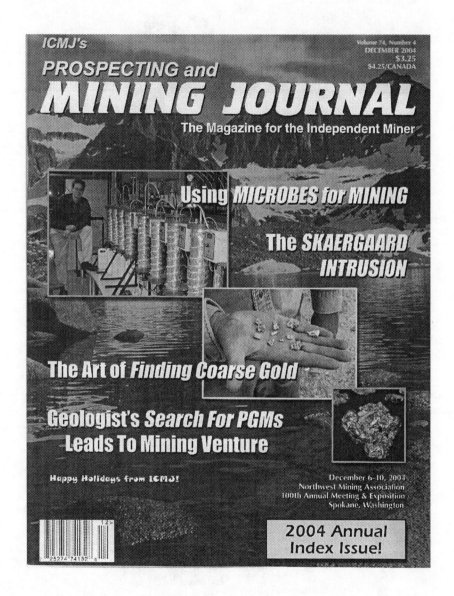

Sean T. Taeschner, M.Ed.

ICMJ's PROSPECTING and MINING JOURNAL
The magazine for the independent miner

PUBLISHER/EDITOR
Scott M. Harn

PUBLISHING ASSISTANT
Desiree Trundy

ADVERTISING/PRODUCTION
Sally Harn
advertising@icmj.com

CIRCULATION/SUBSCRIPTIONS
info@icmj.com

CONTRIBUTING WRITERS
Edgar B. Heylmun, PhD
Leonard Melman
Brian Mooney
Sam E. Phifer, PG
Chris Ralph

AP Associated Press

Direct subscriptions/inquiries/mail to:
ICMJ
PO Box 2260
Aptos, CA 95001-2260
Phone: (831) 479-1500
Fax: (831) 479-4385
E-mail: info@icmj.com
VISIT OUR WEBSITE OR SUBSCRIBE
ON-LINE at: www.icmj.com

ICMJ's Prospecting and Mining Journal (ISSN0008-1299) is published monthly by California Mining Journal, Inc. at 3065 Porter St., #103, Soquel, CA, 95073. US Subscriptions are $25.95 for one year; Canadian subscriptions are $39.50 (U.S. Funds) for one year; and all other international subscriptions are $43.50 (US Funds) for one year. International subscriptions are shipped via air mail. Periodicals postage paid at Aptos, CA and at additional mailing offices. POSTMASTER: Please send address changes to *California Mining Journal*, PO Box 2260, Aptos, CA 95001-2260. *ICMJ's Prospecting and Mining Journal* is purchased with the understanding that information presented herein is from many sources and, while believed to be reliable, ICMJ does not guarantee its accuracy nor assume liability for its use. The publication offers no advice or opinion, and nothing contained herein should be interpreted as our recommendation to buy or sell securities; nor do we offer any type of investment advice. Officers, employees and guest writers may, from time to time, purchase, sell or have a position in securities of companies discussed or mentioned in this publication.
Advertisers' claims are solely their own and there can be no warranty or responsibility by the publisher. ICMJ reserves the right to refuse or discontinue any advertisement if content or performance does not meet editorial policy. ICMJ welcomes articles and photographs of mining operations, mining history and related subjects. Submitted material will be returned, if requested with submission. Copyright 2004

Quality printing was provided by Casey Printing, King City, California. (831) 385-3222.
www.caseyprinting.com

INSIDE

...the procedures to be used with the PGMs vary considerably depending upon the make-up of each ore body.

GEOLOGIST'S SEARCH FOR PGMs LEADS TO MINING VENTURE — **6**

This is the general description of how gold deposits are formed and could be applied to a number of different types of deposits.

THE ART OF FINDING COARSE GOLD—PART I — **14**

...there are applications across many other metals including gold, zinc and nickel.

USING MICROBES FOR MINING — **22**

The intrusion, a mineral-rich volcanic plug embedded in sedimentary rock, is widely regarded as the classic of its kind.

THE SKAERGAARD INTRUSION — **38**

Finding Gold in Washington State: 2005-6 Edition

Northwest Treasure Supply
P.O. Box 4212 Bellingham Wa. 98227-4212
Info line (360) 733-5750 Fax line (360) 733-5758
Order line 1-800-845-5258

Robert "RC" Cunningham

Northwest Distributor for Garrett Products
and Bounty Hunter
Dealer Inqures Welcome
email - nwts@nwtsdetectors.com   www.nwtsdetectors.com

PROSPECTOR'S SUPPLY
A division of Teninsula Sports Cards...2616 Bridgeport Way West; Tacoma, WA
(253) 565-9426

"GRIZZLY" Gold Pan - world's first available pan that lets you remove gold concentrates through the bottom of the pan.

FASTEST - works like a sluice. You can process 3-5 times as much gravel as your old pan.

SAFEST - you will not lose fine gold even when you pan aggressively.

EASIEST pan in the world to learn and to use. A child can be proficient in minutes.

"GOLD MISER" Sluice - world's smallest and most efficient sluice.

Only 10" Long

FASTEST tool for separating fine gold from black sand concentrates.

SAFEST tool for recovering and holding fine gold. Will not lose flour gold.

MOST EFFICIENT tool for testing samples while prospecting. 10-20 times faster than a pan. In seconds a sample is cleaned down to the gold which shows up brightly against the blue color.

Prospecting Book

NEW INFORMATION on how to locate gold concentrations by understanding how gold moves.

NEW EQUIPMENT and old reliable tools - how to use and be more successful no matter where you go.

NEW IDEAS that will help even experienced miners.

OLD IDEAS that are wrong and hurt your chances of getting gold are explained.

Gold Panning Video

GOLD PANNING REINVENTED
- 5-10 Times More Gold
- Science of Gold Panning Explained in Easy Terms
- Easy To Use

GRIZZLY PAN

SLUICE BOX, INC. • 16745 N. Oak Hill Ln. • Mt. Vernon, IL 62864 • (618) 244-0505 • Fax (618) 244-9191
VISIT OUR WEB PAGE www.sluicebox.com

Sean T. Taeschner, M.Ed.

2003 RETAIL PRICES

(ALL PRICES INCLUDE FIRST CLASS SHIPPING AND HANDLING IN THE U. S.)

Gold Panning Kit (includes pan, sluice, book, extra gold pan seal)	$33.95
Deluxe Panning Kit (includes gold panning kit plus video)	$49.95

---

	1 each	2 each
Gold Panning Video (1 hour)	$23.00	$43.00
Grizzly Gold Pan	$15.00	$27.00
Gold Miser Sluice	$11.00	$19.00
Gold Prospecting Book	$9.00	$15.00
Extra Seal for Grizzly Gold Pan	$5.00	$6.50

(Make checks payable to: Sluice Box, Inc.)

GOLD PANNING VIDEO

- Educational and Entertaining
- Helps improve gold recovery by **5-10 times**
- Improves fine gold recovery even for those who continue to use round pans

VISIT OUR WEB PAGE
www.sluicebox.com

SLUICE BOX, INC. • 16745 N. Oak Hill Ln. • Mt. Vernon, IL 62864 • Phone (618)244-0505 • Fax (618)244-9191

Section 10
State of Washington Contact Information

State of Washington
DEPARTMENT OF FISH AND WILDLIFE

Mailing Address: 600 Capitol Way N • Olympia, WA 98501-1091 • TDD (360) 902-2207
Main Office Location: Natural Resources Building • 1111 Washington Street SE • Olympia, WA

March 30, 2005

Sean T. Taeschner
30846 229 PL SE
Black Diamond, WA 98010

Dear Mr. Taeschner:

I received your letter of March 21, 2005 requesting permission to reprint Washington Department of Fish and Wildlife's (WDFW) Gold and Fish pamphlet in your upcoming book, *Finding Gold in Washington State*. WDFW is not able to grant your request.

The Gold and Fish pamphlet is the Hydraulic Project Approval (HPA) issued by WDFW for the most common mineral prospecting activities in Washington State. Because it is a permit, we cannot allow reproductions of the pamphlet to be printed. Reproductions do not satisfy the requirement to have an official copy of the pamphlet on the job site. WDFW does have an unofficial version of the pamphlet on our website. It currently can be found at the following URL: http://wdfw.wa.gov/hab/goldfish/goldfish.htm.

You are welcome to list that website in your book, and to list WDFW contact information that is in the back of the pamphlet so that readers can reach us to request official copies of the pamphlet. Good luck with your book.

Sincerely

Patrick F. Chapman
Habitat Program

Sean T. Taeschner, M.Ed.

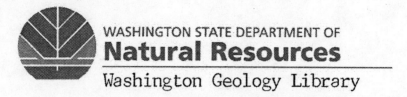

Washington Geology Library

Lee Walkling
Library Information Specialist

GEOLOGY & EARTH RESOURCES
1111 WASHINGTON ST SE
PO BOX 47007
OLYMPIA WA 98504-7007
FAX 360-902-1785
360-902-1473
lee.walkling@wadnr.gov

*NOTE: Reprinted by permission of Washington State Department of Fish & Wildlife and the Washington State Department of Natural Resources Geology Library.

Section 11
Where To Find Gold In Washington State
(Author's Favorite Holes)

- Bear Gap Mine

I was lucky enough to travel with John Davis and his wife and son to explore an old mine about one hour's drive time and ¼ mile in distance from Bear Gap on Highway 410 East out of Enumclaw, Washington yesterday.

Arriving at a hairpin turn by a guardrail some 40 feet west of the Bear Gap sign (it is blue and white)...we drove over a very bumpy dirt road until we came to a clearing right in front of a hole under a large tree. This was the mine entrance. We could see remnants of an old rusted mine car slightly buried under the cave's narrow slit-eye-shaped entrance.

Sean T. Taeschner, M.Ed.

A shard of old railroad rail was visible poking up out of the ground by a large tailings pile some 25 feet from the mine's entrance.

Hunters drove by as they searched to set up camp during hunting season as they sought elk.

The rock on the outside was very hard quartz-iron ore mixture with streaks of small iron pyrite veins running through and dotting the inside of the rocks.

Venturing inside it was clear that there were no bats, rats, bear, or cougars. Some beer bottles littered the mine's floor, as well as rotting log timbers that were once used within the mine. Old wooden slats graced the floor where rails once were fastened. Small drips of water emanated from the ceiling and calcium carbonate and copper sulfate leached from the rocks to form a greenish Comet cleanser-appearing powder. Moths dotted the inside walls…some alive and others dead with mold shards growing off of them like pine needles stabbing out of voodoo dolls.

Using an LED headlamp and hand-held flashlight, John and I ventured some forty feet into the mine. The floor was somewhat wet and level and tunnels branched off into five directions. Using a pick we tried to free some fine pyrite samples and made sparks that would set fear into any gas station attendant filling propane bottles. We did not pass out from any invisible, odorless poisoned gas (Black Damp), did not need any canaries, and did not fear any parts of the ceiling caving in. It was, however, cold inside.

I would not recommend that anyone go into an abandoned mine without a hardhat, good flashlight(s), and a buddy in case of emergency. A gun might also come in handy in case one might run into a bear or cougar.

We did come home, though, with some interesting samples of very bright and flashy silver-streaked rocks. I intend to have them analyzed by geologists to verify their contents.

I had the samples assayed that we found by Jerry's Rock & Gem shop in Renton, Washington. The majority of the samples showed Galena (lead), white quartz, iron, copper, and iron pyrite. No gold or silver were anywhere in the samples.

The owner of Jerry's Rock & Gem, Glen, stated that he had spoken to an old timer from the area that had once mined in the area. The Bear Gap mine had been a lead mine in its heyday.

DISCLAIMER: The author does not condone or suggest ANYONE enter an abandoned mine. Abandoned mines can be extremely dangerous. To do so is at one's own personal risk. The author bears no responsibility whatsoever for the actions of those reading his book and deciding to try it against his advice.

Sean T. Taeschner, M.Ed.

Entrance to the Bear Gap Mine, 2004

- Horseshoe Bend on the Sultan River in the Sultan Basin Watershed...on Highway 2 East of Everett as you head east to Stevens Pass IS THE BEST PLACE TO FIND GOLD. The old timers pulled 80,000 ounces out of there in 1900. They even cut a channel through the bend to divert the river for a year so as to get all the gold...but they didn't get it all! Bedrock is 24 inches down from the river's inside edge and not only is there lots of gold nuggets, but mercury as well from the old timers' gold pans! If signs prohibit prospecting in this area then obey them. They are most likely private property areas or state/federal land(s) or others' claims. Respect is a must in assuring the continuation of our centuries-old hobby. Fill in your holes and pack out yours and others' trash. Don't give someone a reason to shut down this hobby! Be responsible. There are black bears in the area!

Sean T. Taeschner, M.Ed.

Kelly Watson and son at Horseshoe Bend on Sultan River, 1995

Finding Gold in Washington State: 2005-6 Edition

- Goldbar on Highway 2 East of Everett, WA as one drives north on I-5. I go to the public boat launch where the river-rafters put in and the hard-pack and sand are full of gold! You won't be disappointed!

- Icicle Creek about 8 miles southwest of Leavenworth, WA is also a great place to prospect. This is a state campground and I often find more 2 and 3-carat light pink to root beer-brown garnets here when I dig down about two feet. There is some gold as well.

- Gold Creek between I-90 and Hwy 2 at Blewett Pass...any stream coming down from there is full of gold...but heavily claimed...so don't get shot by becoming a claim-jumper. I advise against panning here without written permission from the claim owner(s).

Sean T. Taeschner, M.Ed.

Sean Working a Tub of Muddy Gold at Green River in 2004

Finding Gold in Washington State: 2005-6 Edition

- Liberty, WA, which is North of Peshastin outside of Leavenworth, WA.... still, has open gold mining operations on an industrial scale...and the area is as heavily mineralized as Blewett Pass.

- Green River at Palmer-Kanasket State Park near Enumclaw ...has gold under the rocks and lead as well as fishing hooks, etc. I've often found live rounds in my pan. Maybe a gun range is nearby?

Sean Gold Panning on Green River in 1994

Sean T. Taeschner, M.Ed.

White River Mine Entrance Collapse © 2004

Section 12
Unit & Lesson Plans for Teachers

Sean T. Taeschner, M.Ed.

Gold:

Methods of Discovery & Recovery

(A One-Week Unit Plan for 4th Grade Science)

by "Sluicebox Sean" T. Taeschner

CURRICULUM PRESENTATION

Curriculum Name: Gold: Methods of Discovery & Recovery

Curriculum Grade Level: 4th Grade Science

Curriculum Type: Pro-active (Earth Sciences)

Curriculum Cost: $250.00 to the school district for 25 students (paper, water, bus to museum for field trip, admission tickets)

Curriculum Summary: This unit will last approximately one week (five school days). It will teach 4th graders the importance of gold as a mineral and as a material to be mined for commerce as well as pleasure. Students will be able to relate the contents of this unit to the Washington State 4th Grade Science Benchmark Essential Academic Learning

Requirements (EALRs) and the Kent School District Student Learning Objectives (SLOs). Not only will students learn to find and recover gold but will learn its historical impact on mining history and word usage in the United States of America. The unit will answer the WHO/WHAT/WHERE/WHEN/WHY & HOW to find, refine and profit from recreational and commercial gold prospecting.

The unit will begin with students being curious about what a gold pan is and culminate with a field trip by school bus to the Klondike Gold Rush Museum in Pioneer Square in Seattle.

Statement of Need and Impact: Students need to learn that gold is a discoverable and recoverable metal, which will excite them in the field of mining in earth science studies.

Students will be impacted in their daily lives historically, recreationally and commercially as they discover the many uses of gold.

Planning Process

Gold: Methods of Discovery & Recovery

(A 5-Day Unit Plan for 4th Grade Science)

by "Sluicebox Sean" T. Taeschner

STAGE ONE

1) Enduring Understandings Desired
   Students will understand:
   a) Washington State EALRs
      (5th Grade Science - Benchmark (1,1.1.1,1.1.4,1.2,1.3)

    b) Kent School District #415 SLOs (1.0, 2.0, 3.0)
    c) Gold prospecting as a hobby can teach life-long basic mining skills, which can be used to get a job as an adult working in a real gold mine.
    d) Practical earth science lessons in school can help students move from 'Why do we have to learn this stuff?' to 'I can use this to feed my family!'

2) The Essential Questions To Guide The Unit And Focus On Teaching And Learning
    a) Why do men get 'gold fever'?
    b) Is America really 'paved with gold'?
    c) Is learning earth science important for helping you make a living doing this?

3) Key Knowledge And Skills Students Will Acquire As A Result Of This Unit?

Students Will Know:
a) Basic gold identification
b) How to process the concentrates
c) How to explain what the gold cycle is

Students Will Be Able To:
a) Read a river
b) Pan for gold
c) Recover / collect their gold
d) Explain and demonstrate simple gold recovery methods

STAGE TWO
Determining Acceptable Evidence
Performance Tasks

1) Written: Students complete a bubble map of what gold is and write in a reflection journal about what gold prospecting is.
2) Oral: Students explain what the 'gold cycle' is in earth science terms.
3) Performed: Students pan for gold in class.

Quizzes, Tests, and Prompts:

1) Students answer in a one page essay 'Is this important to learn and could I get a job doing it someday?'
2) Students break into small discussion groups to answer these questions and seek insight from each other.

Unprompted Evidence:

1) The teacher walks the room to observe which students can readily recognize gold versus black sand versus Fools Gold, lead, or garnets as they 'pan for gold' in class.
2) The teacher asks himself/herself 'Are they excited about the discovery process?'

Self-Assessment:

1) The teacher asks himself/herself, "Shall I have ELL learners (with no grasp yet on English) or severely handicapped students (needing suctioning, diaper

changing, etc.) draw a picture of gold miners instead? Or, do they understand and shall I allow them to go along with the class and attend the final field trip or have a substitute teacher watch them while we are away?"

2) The teacher asks himself/herself, "What do I need to change about the unit or lessons the next time I teach it? What went well? What did not?"

3) The teacher asks himself/herself, "How do I check for enduring lifelong understanding in the students? Bubble maps at end of unit? Shall I have the students fill out an evaluation sheet on my teaching method(s) or teaching style for the unit? Suggestions for change from them?"

STAGE THREE
Plan Learning Experiences And Instruction

Materials Needed: 10 gold pans, gold bearing soil, black sand, garnets, lead, gold, iron pyrite (Fools Gold), water, three washtubs, sluice box, rocker box, 4th grade science book, school bus, $250.00 for 26 museum tickets.

Order of Learning Experiences (Unit Schedule)
DAY ONE: Monday- (The Big Prop) with Guessing Jar on table as students come in (Sparks curiosity and discovery in the students. Sign with 'What Is This?" Fill out the card and drop it into the jar. Prize drawing is this Friday.)
- Teacher pans for gold in class as students watch and he/she lectures about the concept of recreational gold prospecting.

- Students write in reflection journal or small Flip Book about what they think gold prospecting is.

DAY TWO: Tuesday-The teacher asks what we learned yesterday? Explain to students what we will learn today.
- The teacher presents commercial prospecting and explains how to identify gold and other items in a gold pan
- (Oral presentation) The students volunteer to explain what the 'gold cycle' is from their earth science books.
- The students break up into groups of four and pan for gold during class.

DAY THREE: Wednesday-The teacher asks students what we learned yesterday and explains what we will learn today.
- Students get to learn about gold prospecting history by watching a video "U.S. Gold Discovery: The Gold Rush"

(No assigned tasks or homework. The teacher will allow this all to absorb into the students' minds)

DAY FOUR: Thursday-The teacher asks students what they saw in the video yesterday and goes over the field trip rules for safety for tomorrow.
- The teacher asks the students, "Can we get a job by learning this stuff?" This stimulates discussion.
- Students break up into small groups of four to discuss this question and write a small one-page essay answering it. They then share it with each other.

- The students are asked to come to a front table to identify gold, iron pyrite, black sand, garnets, lead and other quartz rocks on individual pieces of paper, which are folded and placed into a prize jar for Friday (Second chance at winning-if they have the correct answer).

This is a learning assessment tool for the teacher to assess their understanding of the unit so far and add or change anything before it is over tomorrow.

- The teacher also wanders the room and observes students as they discuss and identify the key questions and answers to the written and visual stimulants in today's lesson.

DAY FIVE: Friday- (Four Hour Field Trip)
- The teacher asks students, "What was it that they discussed yesterday about making a living at mining for gold. Is it possible to make a living at mining it?"
- The teacher takes the students on a field trip to the Klondike Gold Rush Museum in Pioneer Square in Seattle.
- After returning from the trip, the teacher holds the prize drawing (A new gold pan is given away) for the students.

Resources

1) Taeschner, Sean T. "Finding Gold In Washington State" © 1999.

2) Klondike Gold Rush Museum – Pioneer Square, Seattle, WA USA.

Finding Gold in Washington State: 2005-6 Edition

3) "U.S. Gold Discovery: The Gold Rush" (A 4 part video series by The History Channel) © 2002.

4) "Gold and Fish" – Washington State Department of Fisheries © 1995.

5) Washington State (5th Grade Benchmark) Earth Science EALRS.

6) Kent School District #415 SLOs (Student Learning Objectives).

7) (Pretend) 4th Grade Science Textbook (Earth Sciences Section).

8) World Wide Web (Internet) Search Engines:

   a. http://www.Google.com
   b. http://www.webquest.sdsu.edu

<u>WebQuest Examples: Grades 3-5 Social Studies</u> •
Grades 3-5 Social Studies WebQuests = newly listed since 4/1/2002 "A Forest Forever." Decide the fate of a newly designated National Forest. A...

...And They Came to the Streets That Were Paved With Gold Chinese immigration to California during the Gold Rush-1882 Art for Sale Look at art and...

...Eureka! Gold Rush Journal (Australian) Write a four week journal describing your trip to an Australian gold field and your experiences there sometime...

Sean T. Taeschner, M.Ed.

...Gold Rush Players! Write a Gold Rush play based upon historical research. Grandpa's Mountain Uses historical fiction for understanding issues...
36% Thu, 25 Jul 2002 06:02:13 GMT
http://webquest.sdsu.edu/matrix/3-5-Soc.htm

NOTE: Thanks to the Kent School District #415 and the Office of Superintendent of Public Instruction in Olympia, WA for allowing me to republish and use their EALRS and SLOs.

Gold: Methods of Discovery & Recovery

A WebQuest for 4th Grade Science (Gold:
Methods of Discovery & Recovery)
A 5 day unit plan designed by Sean T. Taeschner
Sean T. Taeschner
STeshner@Juno.com

Introduction | Task | Process | Evaluation | Conclusion | Credits | Teacher Info

Finding Gold in Washington State: 2005-6 Edition

Introduction

You are a weekend gold prospector trying to identify the location of the glory hole in Palmer-Kanasket State Park. But first, you need to learn a few methods for discovering gold in that glory hole before you can collect and refine it.

Do you think that learning these skills could enable you to make a living at what you really enjoy doing…being outdoors… and encouraging others to enrich America by becoming a gold miner? Let's find out!

The Task

As a gold prospector looking for that elusive glory hole, you will become an active learner in discovering that gold is a discoverable and recoverable metal. Will you be excited about how this field of mining in earth science studies could make you rich someday or help you to learn a real-life skill to feed your family with?

You will be impacted as never before as an active participant and peer coach in your daily life from this day forward, both historically, recreationally and commercially while discovering the many uses of gold! Now, let's see if things pan out as described!

You will work with your fellow students to create a special Double-Bubble gold map using Inspiration Software http://www.kent.k12.wa.us/staff/ksd14526/inspiration/inspiration.htm and by searching about Gold Prospecting on a search engine called www.Google.com to help you in your prospecting journey. You will actively discuss what you have learned and discovered, such as panning for gold and telling others how to find it. Let's dig in!

Product to be designed:

Students will complete a bubble map of what gold is and write in a reflection journal about how gold prospecting could enable them to make a living as gold miners. Students will then discuss what they have learned in small groups with their peers.

Students will look up information for their reflection papers by using websites.

(Directions to Students) "You will access at least two websites to look up information on gold prospecting for your reflection papers. Click on the website hyperlinks below that are in blue or purple." WebQuest Examples: Grades 3-5 Social Studies and http://www.Google.com

"You will also use the following link in Inspiration Software to create your Double Bubble map: http://www.kent.k12.wa.us/staff/ksd14526/inspiration/inspiration.htm

The final product will involve using Inspiration, and Microsoft Word."

The Process

Student Performance Tasks

1) Written: Students complete a bubble map of what gold is and write in a reflection journal about what gold prospecting is. Could they use these skills to get a job someday working as gold miners? They will use a double-bubble map format using Inspiration Software at the following web-link: http://www.kent.

k12.wa.us/staff/ksd14526/inspiration/inspiration.htm

2) Oral: Students explain what the 'gold cycle' is in earth science terms.
3) Performed: Students pan for gold in class.

Quizzes, Tests, and Prompts

1) Students answer in a one page essay: 'Is this important to learn and could I get a job doing it someday?'
2) Students break into small discussion groups to answer these questions and seek insight from each other.

Non-Direct Teacher Tasks

1) The teacher walks the room to observe which students can readily recognize gold versus black sand versus Fools Gold, lead, or garnets as they 'pan for gold' in class.
2) The teacher asks himself/herself 'Are they excited about the discovery process?'

Plan Learning Experiences And Instruction

Materials Needed: 10 gold pans, gold bearing soil, black sand, garnets, lead, gold, iron pyrite (Fools Gold), water, 3 washtubs, sluice box, rocker box, 4th grade science book, school bus, $250.00 for 26 museum tickets.

Order of Learning Experiences (Unit Schedule)

DAY ONE: Monday- (The Big Prop) with Guessing Jar on table as students come in (Sparks curiosity and discovery in the students. Sign with 'What Is This?" Fill out the card and drop it into the jar. Prize drawing is this Friday.)
- Teacher pans for gold in class as students watch and he/she lectures about the concept of recreational gold prospecting.
- Students write in reflection journal or small Flip Book about what they think gold prospecting is.

DAY TWO: Tuesday-The teacher asks what we learned yesterday? Explain to students what we will learn today.
- The teacher presents commercial prospecting and explains how to identify gold and other items in a gold pan
- (Oral presentation) The students volunteer to explain what the 'gold cycle' is from their earth science books.
- The students break up into groups of four and pan for gold during class.

DAY THREE: Wednesday-The teacher asks students what we learned yesterday and explains what we will learn today.
- Students get to learn about gold prospecting history by watching a video "U.S. Gold Discovery: The Gold Rush."

(No assigned tasks or homework. The teacher will allow this all to absorb into the students' minds.)

DAY FOUR: Thursday-The teacher asks students what they saw in the video yesterday and goes over the field trip rules for safety for tomorrow.

- The teacher asks the students, "Can we get a job by learning this stuff?" This stimulates discussion.
- Students break up into small groups of four to discuss this question and write a small one-page essay answering it. They then share it with each other.
- The students are asked to come to a front table to identify gold, iron pyrite, black sand, garnets, lead and other quartz rocks on individual pieces of paper, which are folded and placed into a prize jar for Friday (Second chance at winning-if they have the correct answer).

This is a learning assessment tool for the teacher to assess their understanding of the unit so far and add or change anything before it is over tomorrow.

- The teacher also wanders the room and observes students as they discuss and identify the key question and answers to the written and visual stimulants in today's lesson.

DAY FIVE: Friday- (Four Hour Field Trip)
- The teacher asks students, "What was it that they discussed yesterday about making a living at mining for gold. Is it possible to make a living at mining it?"
- The teacher takes the students on a field trip to the Klondike Gold Rush Museum in Pioneer Square in Seattle.
- After returning from the trip, the teacher holds the prize drawing (A new gold pan is given away) for the students.

Sean T. Taeschner, M.Ed.

Evaluation (Grading Rubric)

Performance will be evaluated for a common grade for group work and individual work as follows:

Each white area= 5 points	Beginning 25 pts max	Developing 25 pts max	Accomplished 25 pts max	Exemplary 25 pts max	Score 100%
Written: Students complete a bubble map of what gold is and write in a reflection journal about what gold prospecting is. Could they use these skills to get a job someday working as a gold miner?	Double-Bubble Map is handed in.	Double-Bubble Map is made on a PC.	Double-Bubble Map is made on a PC in full color, done accurately, and is descriptive. The reflection journal is turned in.	Double-Bubble Map is made on a PC in full color, done accurately, is descriptive, and compares & contrasts gold to Fools Gold, garnets, and lead. Skills to get a job someday working as a gold miner are included in the reflection journal on gold prospecting.	

Oral: Students explain what the 'gold cycle' is in earth science terms.	Student understands that gold can be recycled like water or rocks or air.	Student understands that gold can be recycled like water or rocks or air and can explain it on paper.	Student understands that gold can be recycled like water or rocks or air and can explain it on paper and out loud to classmates.	Student understands that gold can be recycled like water or rocks or air, can explain it orally with peers and in writing and can teach it to the class.	
Performed: Students pan for gold in class.	Student tries to pan for gold.	Students pan for gold and can separate black sand from gold in the pan.	Students pan for gold and can separate black sand from gold in the pan, as well as identify the difference between fools gold and real gold in the pan.	Students pan for gold and can separate black sand from gold in the pan, as well as identify the difference between Fools Gold and real gold in the pan and teach peers about their discovery.	

Students answer in a 1 page essay: 'Is this important to learn and could I get a job doing it someday?'	Student does not hand in the essay.	Student hands in the essay but does not explain what he/she has learned.	Student hands in a completed essay that covers the main points learned in class. The paper is handwritten, hastily completed, and lacks detail.	Student hands in a completed essay, which is completed on a word processor, includes graphic organizers, labels, pictures, or charts, and sites resources for information beyond main points learned in class. The paper shows true scholarship.	
Students break into small discussion groups to answer these questions and seek insight from each other.	The student did not participate at all in the group discussion and no reflection paper was handed in.	There was some participation but no interest in learning of the topic. The reflection paper was incomplete.	Learning was demonstrated by active participation in the group discussion and completion of a reflection paper afterwards.	The student demonstrated enthusiasm for the topic discussed in small group discussion and helped peers further understand what was taught for enduring understanding. The reflection paper handed in showed true scholarship and was an example to peers of excellence.	

Conclusion

Key Knowledge And Skills Students Will Acquire As A Result Of This Unit

Students Will Know:
   a) Basic gold identification
   b) How to process the concentrates
   c) How to explain what the gold cycle is

Students Will Be Able To:
   a) Read a river
   b) Pan for gold
   c) Recover / collect their gold
   d) Explain and demonstrate simple gold recovery methods

Credits & References

1) Taeschner, Sean T. "Finding Gold In Washington State" © 1999. http://www.amazon.com/exec/obidos/tg/detail/-/0970843305/qid=1049684321/sr=1-2/ref=sr_1_2/002-7688475-1520061?v=glance&s=books

2) Klondike Gold Rush Museum – Pioneer Square, Seattle, WA USA.

3) "U.S. Gold Discovery: The Gold Rush" (A four part video series by The History Channel) © 2002.

4) "Gold and Fish" – Washington State Department of Fisheries © 1995.

5) Washington State (5th. Grade Benchmark) Earth Science EALRS.

(5th Grade Science- Benchmark (1,1.1.1,1.1.4,1.2,1.3)

6) Kent School District #415 SLOs (Student Learning Objectives).

(1.0, 2.0, 3.0)

7) (Pretend) 4th. Grade Science Textbook (Earth Sciences Section).

8) World Wide Web (Internet) Search Engines:

   a. http://www.Google.com
   b. http://www.webquest.sdsu.edu

1. WebQuest Examples: Grades 3-5 Social Studies •
Grades 3-5 Social Studies WebQuests = newly listed since 4/1/2002 A Forest Forever. Decide the fate of a newly designated National Forest. A...

...And They Came to the Streets That Were Paved With Gold Chinese immigration to California during the Gold Rush-1882 Art for Sale Look at art and...

...Eureka! Gold Rush Journal (Australian) Write a four week journal describing your trip to an Australian gold field and your experiences there sometime...

...Gold Rush Players! Write a Gold Rush play based upon historical research. Grandpa's Mountain Uses historical fiction for understanding issues...

Finding Gold in Washington State: 2005-6 Edition

36% Thu, 25 Jul 2002 06:02:13 GMT http://webquest.sdsu.edu/matrix/3-5-Soc.htm

Curriculum Presentation

Curriculum Name: Gold: Methods of Discovery & Recovery

Curriculum Grade Level: 4th Grade Science

Curriculum Type: Pro-active (Earth Sciences)

Curriculum Cost: $250.00 to the school district for 25 students (paper, water, bus to museum for field trip, admission tickets)

Curriculum Summary: This unit will last approximately one week (five school days). It will teach 4th graders the importance of gold as a mineral and as a material to be mined for commerce as well as pleasure. Students will be able to relate the contents of this unit to the Washington State 4th Grade Science Benchmark Essential Academic Learning Requirements (EALRs) and the Kent School District Student Learning Objectives (SLOs). Not only will students learn to find and recover gold but will learn its historical impact on mining history and word usage in the United States of America. The unit will answer the WHO/WHAT/WHERE/WHEN/WHY & HOW to find, refine and profit from recreational and commercial gold prospecting.

Sean T. Taeschner, M.Ed.

The unit will begin with students being curious about what a gold pan is and culminate with a field trip by school bus to the Klondike Gold Rush Museum in Pioneer Square in Seattle.

Enduring Understandings Desired

Students will understand:
1) Washington State EALRs
(5th Grade Science- Benchmark (1,1.1.1,1.1.4,1.2,1.3)
2) Kent School District #415 SLOs (1.0, 2.0, 3.0)
3) Gold prospecting as a hobby can teach life-long basic mining skills, which can be used to get a job as an adult working in a real gold mine.
4) Practical earth science lessons in school can help students move from 'Why we have to learn this stuff?' to 'I can use this to feed my family!'
5) The Essential Questions To Guide The Unit And Focus On Teaching And Learning
   a) Why do men get 'gold fever'?
   b) Is America really 'paved with gold'?

Is learning earth science important for helping you make a living doing this?

Teacher Info

Sean T. Taeschner, M. Ed. is a native of the Pacific Northwest and has lived in Washington State since 1964. His weekends are spent teaching others how to prospect for gold on the Green River near his home in historic Black Diamond of turn-of-the-century coal mining fame. A graduate of Western Washington University in German language, literature and culture, teacher, and a self-employed remodeling contractor,

Finding Gold in Washington State: 2005-6 Edition

Sean has also authored other books; *Marshmallows With Monica, Finding Gold In Washington State, Finding Gold In Oregon, & Before You Buy A Car...Dirty Dealer Finance Tricks* as well as a music CD, *Hot Smokey Burnout* with his twin brother, Mark, who is an electrical engineer doing flight test instrumentation at Boeing Field in Seattle.

Based on a template from The WebQuest Page

NOTE: Thanks to the Kent School District #415 and the Office of Superintendent of Public Instruction in Olympia, WA for allowing me to republish and use their EALRS and SLOs.

Lesson Plan

Sean T.Taeschner
March 09, 2003
4th Grade Science
One – one hour lesson

Preparation

Objective: Students will understand how to use a gold pan by correctly stratifying the material in it to successfully recover gold.

Sean T. Taeschner, M.Ed.

EALR:

> Essential Academic Learning Requirements—Science
>
> Benchmark 1-Grade 5
>
> 1. The student understands and uses scientific concepts and principles.
>
> To meet this standard, the student will:
>
> 1.1. Use properties to identify, describe, and categorize substances, materials, and objects, and use characteristics to categorize living things.
>
> 1.2.
>
> 1.1.1 PHYSICAL SCIENCE
>
> Properties of Substances: Use properties to sort natural and manufactured materials and objects, for example, size, weight, shape, color, texture, and hardness.
>
> 1.1.4 EARTH/SPACE SCIENCE
>
> Nature and Properties of Earth Materials
>
> Observe and examine physical properties of earth materials, such as rocks and soil, water (as liquid, solid, and vapor) and the gases of the atmosphere.
>
> Benchmark 2—Grade 8
>
> Classify rocks and soils into groups based on their chemical and physical properties; describe the processes by which rocks and soils are formed.
>
> 1.2. Recognize the components, structure, and organization of systems and the interconnections within and among them.
>
> 1.3 EARTH/SPACE SCIENCE
>
> Processes and Interactions in the Earth System
>
> Identify processes that slowly change the surface of the earth such as erosion and weathering, and those that rapidly change the surface of the earth, such as landslides, volcanic eruptions, and earthquakes.

Kent SLO:

Utilizing the Scientific Method to Explore
Universal Connections
Grade Five

1.0 ...determine the importance of water to the universe using experiments through projects, charts, observations, diagrams, models, discussion, and/or written composition.

2.0 ... to illustrate cause and effect of simple machines, to develop solutions using inquiry through models, charts, diagrams, and/or technology.

3.0 ...demonstrate understanding of the structures of the water cycle, forces and motion, and soil erosion, and soil composition through models, collections, illustrations, and/or diagrams.

Materials:

One 12 inch green plastic gold pan with Chinese riffles
One dishpan full of clear water
Five cups of mineralized riverbank soil with lead nuggets
One classroom desk large enough to seat seven students and one teacher
Paper towels to dry student and teacher hands

Introduction (3 minutes)

Anticipatory Set: The teacher will gain student attention by smiling and pouring several cups of riverbank soil into a gold pan, will pan it to stratify the material in it, then pick out a nugget and yell excitedly, "Holy Jehosifer! Gold! Yeller Gold!" He will then hold it in front of students' faces within inches and ask, "Got the fever yet?" He will then ask students to volunteer to answer the following question: Where does gold come from and how does one recover it and why use a pan to do it? Holding the pan up, he will ask if any students know what the dents (Chinese riffles) in the pan are for?

Communication of Purpose: "Today we will be learning how the dents ended up in gold pans and what they were used for and why they are still used in gold prospecting with gold pans today, and will learn how to stratify the material in our gold pans using water, simple riverbank soil, lead nuggets, a dishpan, and a little patience. Then, after you know how to separate out gold from riverbank soil, you will be ready to go out and prospect for gold in a nearby stream or river."

Body of Lesson

(Learning Strategies)

Presentation/ Instruction: (30 minutes)

- The teacher will orally explain to the students that gold has a specific gravity of 19 on the Periodic Table of the Elements and that it hides easily in soil in rivers.
- The teacher will explain that gold miners have perfected the technique of separating or 'classifying' the gold away from the remaining materials during recovery on a large and small scale over several hundred years.
- The teacher will explain how water is used as the cheapest and easiest form of gold separation and

recovery: from commercial dredge, to high-banker sluice-box or to the smallest form of separation, the use of a gold pan.
- Students will follow along with the demonstration by actively listening and writing down questions they might have for later discussion.
- Students will work individually or with a partner at the demonstration table practicing holding and swirling their gold pan and material in the dishpan provided. NOTE: They will have been encouraged by the teacher to help one another stratify the riverbank soils in the gold pan if one student didn't quite master the technique of swirling or dipping the pan in the dishpan's water.
- Once students have mastered the technique of separating the gold nuggets (lead nuggets) out of the gold pan they will take their seats and allow peers to take a turn.
- Students will wait quietly for the question/ answer period at the lesson's end.
- The teacher will work with any special needs students or English language learning students in re-demonstrating gold panning and allow extra practice time if needed.
(10 minutes)
- The teacher will call on students to answer any questions about today's lesson.
(10 minutes)
- The teacher will ask students if today's lesson could help them make money at gold prospecting in their

futures? And, is this a hobby or serious business to be considered in today's world economy?
- The teacher will assign a one page student reflection essay based on what they feel they learned in today's lesson for their homework.
- Students will be allowed remaining class time to work on their assignments quietly.
- During remaining quiet time the teacher will clean up the demonstration table with the aid of student volunteers.

Processing (Diverse Learners/ Cultural/ Linguistics

- Students will process the information communicated in this lesson visually as each step of gold and soil stratification is demonstrated, through actively and tactilely practicing riverbank soil stratification independently or with a partner.
- Students with special needs or who are English language learners will remain as observers if they fear participating or the teacher is concerned that they might eat the material samples such as the lead nuggets. Extra time during the lesson or partner assistance will be given to assist such students.
- Throughout the lesson students will be using active participation and cooperative learning to discover how gold is recovered from riverbank soils while using problem solving and critical thinking.
- Students will use critical thinking skills while describing how gold is separated and recovered from riverbank soils in a written one-page reflection

essay (describe what you learned today) homework assignment.
- Social growth will be developed through partner work.
- Intellectual growth will develop during the entire lesson with students actively participating as individuals and in cooperative learning partnerships.

Monitoring/ Check for Understanding (Classroom Management):

- The teacher will ask students during the introductory demonstration to raise their hands and answer the question of where the gold pan riffles originated from in order to assess if group members are actively listening and interested in the lesson.
- During guided practice, the teacher will wander around the demonstration table to monitor whether the students are able to stratify the soil and gold materials correctly. If not, he will help them practice holding the pan and swirling it in the dishpan water correctly until nuggets surface and students smile with gold fever.
- The teacher will make himself available during independent and cooperative learning guided practice and student question/ answer times.

Communication:

- The teacher will use <u>verbal communication</u> in starting the lesson to gain student attention using the phrase, "Eyes on me…freeze!"

- The teacher will encourage students to raise their hands if they need to use the restroom, but will ask them at the beginning of the lesson to write their questions down on paper to be answered at a later period during the lesson prior to closure.
- The teacher will wander around the students and use <u>close physical proximity</u> to monitor behavior and learning and encourage those who are not on task to get on task.

Closing (7 minutes)

Closure: (Points to Ponder) 4 minutes

- Who thinks that using water to separate gold from riverbank soils is the best and cheapest way to prospect for gold? Is it environmentally sound?
- What strategy have we learned for separating gold from heavy and lightweight soils in rivers? How does water and gravity work to help us do this?
- Where is gold found in the United States today? Is there any left to mine? Prospect for?
- When is the best time of the year to dredge or pan for gold?
- Why do water and gravity and the weight of gold itself play an important part in its recovery?
- How would you teach a friend to prospect for and find gold?
- Could you make any money at prospecting for gold today in the USA?

Assessment:(2 minutes)

- I will call on students during the closing part of my lesson to volunteer to tell the class what they learned today about gold panning. (See closure above)
- I will assess student learning in today's lesson by having students write a one-page reflection essay on what they learned about soil classification, gravity and water, and the weight of gold all affecting its recovery.
- I will remind them that the assignment is due the following school day and should be single-spaced and written in complete sentences.

Follow-Up: (1 minute)

- Students will be encouraged by their teacher to ask a parent or adult friend to go with them to a river to try to prospect for gold using a gold pan, rocker box, or sluice-box.
- Students will be asked by their teacher to consider asking the 2nd grade teacher if they can hold a similar in-class gold panning demonstration for the students in their school. Could they relate it to pirates and buried treasure stories?

Reflection:

- Students will use their newly-honed gold panning skills to look for gold, garnets, arrowheads, and industrial diamonds while recreational prospecting or as mining engineers in the future.

- Evidence of learning will be assessed by reading and grading their one-page reflection essays and by looking at the gold they bring into my classroom after weekend prospecting trips.
- If I were to teach this lesson again I would include garnets and fishing tackle as other items, which often can show up in a prospector's gold pan. I would also warn them about the dangers of panning in clay (it sticks to the gold and disappears from the gold pan or sluice box easily) and ask them to fill in their gold prospecting holes, pack their garbage out, and wear life jackets by the rivers they prospect in. Unfortunately, gold prospecting could be an entire week's unit since there are so many aspects of gold mining that have nothing at all to do with the use of water in the recovery process. I could expand on the many ideas of gold separation both commercially and as a hobby and gold recycling, smelting, and mining.

NOTE: Thanks to the Kent School District #415 and the Office of Superintendent of Public Instruction in Olympia, WA for allowing me to republish and use their EALRS and SLOs.

Section 13
Disclaimer

This is an informational book only. No part of this book suggests doing something illegal or dangerous in the hobby of gold prospecting. The author in no way suggests doing any claim jumping, breaking any local or state laws regarding private or public property (Get written permission from the owner(s) or park ranger(s) before attempting to prospect), using cyanide or mercury or nitric acid to separate gold. It is against the law to melt down your gold. Only the U.S. Mint can do this or licensed jewelers. Entering old abandoned mines is EXTREMELY DANGEROUS and should not be attempted at all by anyone except a professional.

If you bought the CD, or diskette it is for IBM Version computers only and must be viewed and printed using Microsoft

Sean T. Taeschner, M.Ed.

Windows and MS Word for Windows. No part or whole may be copied without written permission from the author. Such media is shipped virus-free and use of the media, once the packaging is opened, is at the buyer's own risk. The author is not responsible for any damage caused to a computer(s), or DVD player, by usage in either viewed or printed format.

Section 14
Questions?

E-mail: STeshner@Juno.com or Trashner@Hotmail.com

or write to:
Sean T.Taeschner, M.Ed.
30846 229 PL SE
Black Diamond, WA 98010
USA
or call me at (360) 886-1262

Section 15
Acknowledgements

The author would like to thank the following organizations and people who contributed to this book:

Betty & Russ Nation of Jade Drive Rock Shop in Shelton, Washington

Bob "Stiff Neck" Vertefeuille of Bremerton, Washington

Bud Neale of Gold Claimer & Feed Conveyors in Oregon

Carl Pederson of North Central Washington Prospectors in Wenatchee, Washington

Charles Burpee of Enumclaw, Washington

Clarence "Doc" Ashcroft of Sluice Box, Inc. in Mt. Vernon, Illinois

Dave Rutan of Oregon Gold Trips in Grants Pass, Oregon

Honcoop Highbankers

Sean T. Taeschner, M.Ed.

John Davis of Enumclaw, Washington
Kent School District #415 in Kent, Washington, USA (for the use of their SLOs-Student Learning Objectives)
Mark Erickson of Resources Coalition
Robert "RC" Cunningham of Northwest Treasure Supply in Bellingham, Washington
Scott M. Harn of ICMJ's Prospecting and Mining Journal (Magazine of the Independent Miner)
Tom Bohmker, Author of "The Elusive Pocket Gold of Southwestern Oregon"
Vic Pisoni, Moderator of Northwest Underground Explorations, Seattle, Washington

WASHINGTON PROSPECTORS MINING ASSOCIATION in Seattle, Washington
Washington State Department of Fish & Wildlife c/o Patrick F. Chapman, Olympia, Washington
Washington State Department of Geology-Library Section c/o Lee Walkling, Olympia, Washington
Washington State Office of the Superintendent of Public Instruction (for the use of their EALRs-Essential Academic Learning Requirements), Olympia, Washington

About the Author

Sean T. Taeschner, M.Ed. is a native of the Pacific Northwest and has lived in Washington State since 1964. During the school week he teaches middle school English until he can complete his weekends teaching others how to prospect for gold on the Green River near his home in historic Black Diamond, a town of turn-of-the-century coal mining fame. A graduate of Western Washington University in German language, and Eastern Washington University in Elementary Education, Sean has authored several books. Sean's works can be purchased at http://www.Amazon.com and http://www.GhostTownsUSA.Com.

CPSIA information can be obtained at www.ICGtesting.com
Printed in the USA
LVOW080549041212

309987LV00001B/85/A